Studies in Geometry Series

Constructions
Workbook

Creating Geometric Figures

Tammy Pelli

A Breath of Fresh Air

GarlicPress

Published by:
Garlic Press
605 Powers St.
Eugene, OR 97402

ISBN 1-930820-43-7
Order Number GP-143

www.garlicpress.com

Table of Contents

Introduction

- Geometry is about understanding shapes. One of the best ways to develop that understanding is to draw them using the tools of geometry: a compass and a straightedge. The purpose of this workbook is to explore the relationships between lines, curves, and angles by drawing geometrical figures.

- Because every textbook is different, reference to specific theorems and postulates is not made overtly so that this workbook can complement a student's study of geometry without conflicting with his/her schoolwork. Many theorems, postulates, and definitions will be discovered and developed through the activities.

- Towards the end of the workbook, you will find an Exam so that you can check to find if you have mastered the concepts presented here.

- In order to give instructions for drawing the figures, basic geometrical vocabulary is used. A brief Glossary of vocabulary and concepts is found at the end of this workbook, preceding the Answer Key.

- The Answer Key provides the answers to all the practice exercises. The constructions are shown and explanations are made.

Chapter 1

Constructions

A Beginning

What is a construction?

In geometry there is a difference between drawing a geometric picture and doing a geometric construction. The main difference is in the tools used for each method. Drawings allow any tools to be used: pens, pencils, rulers, protractors, compasses, etc. *Constructions require that only a pencil, a straightedge (which can be a ruler) and a compass be used.*

What can be constructed?

Line segments, arcs, circles, triangles, polygons, parallel lines, perpendicular lines…any geometric shape can be created using only a compass, a straightedge and a pencil by applying the basic construction methods you will learn in this book.

Using a compass

A compass has two arms which are joined at the top. One arm ends in a point and the other arm holds a pencil. The arms can be adjusted so that the distance between the point and the pencil is bigger or smaller. When drawing with a compass it is important that you hold the compass in one of these ways:

1. Place the point of the compass on the paper. Then hold only the small part above where the arms meet to spin the compass and to draw an arc or circle.

2. Place the point of the compass on the paper. Then turn the compass by holding the arm that has the point of the compass and let the pencil move over the paper to draw the arc or circle.

3. Place the point of the compass on the paper. Then hold only the small part above where the arms meet so that both the point and the pencil softly touch the paper. Then use your other hand to slowly turn the paper underneath the compass so that the pencil can trace out the arc, or circle, on the moving paper.

What do line segments, arcs and angles look like?

A line segment has two endpoints that are connected by a straight line. An arc is part of a circle. An angle is two line segments that are joined together at one of their endpoints.

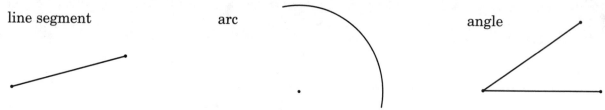

line segment   arc   angle

What does it mean to "measure" without a ruler or a protractor?

When the instructions for a construction include the word "measure," this requires the use of a compass. To measure the length of a segment, put the point of the compass on one end of the segment and stretch the compass until the pencil point meets the other end of the segment. To measure an arc, put the point of the compass at one intersection point on the arc and stretch the compass until the pencil point meets the other intersection point on the arc. *Compass width* is how far apart the pencil point and the point on the other arm of the compass are. When you measure the length of a segment or measure an arc, the measurement is the compass width. In constructions, *measure* never means to use a ruler to measure length, nor does it mean to use a protractor to measure an angle.

What does it mean to intersect?

When 2 lines intersect, they cross each other (think of the intersection where two streets meet). An arc and a circle can intersect if they meet. Intersections are the points where two objects cross each other.

Following directions matters.

Once you learn the basics of doing constructions, you will see similarities in what you are asked to do in each set of instructions. Here are some common directions you will be given and examples of what the results can look like. Keep in mind that everyone's construction looks a bit different, and these are merely guidelines to give you a sense of whether or not you are on the right track.

Draw a line with the straightedge.

Draw a segment and show the endpoints clearly. (Don't make them too big or your construction will not be precise.)

Put the compass on the endpoint and make a small arc that intersects the segment.

Draw an arc that intersects both sides of the angle.

Congruent objects

This is an important word in geometry, and in constructions. Congruent is similar to equal. When two segments are equal, they have the same length as expressed in inches or centimeters or some other unit of measurement. *Congruent* simply means that two objects have the same size and shape. So, two congruent segments would have equal lengths. The word congruent is helpful in geometry because we can say that two triangles are congruent (or any two geometric objects) as a short-cut for saying that all three sides and angles of one of the triangles have the same measures as the corresponding sides and angles of the other triangle.

How do I use this book?

As you read the instructions for how to do a construction, try to follow along with the example. For instance, if it says to draw an arc, trace the arc that has been drawn in the example with your compass. This will give you a feel for how the construction is done. Then when you try it for yourself, you will already have gone through the motions once. You might even want to try the example on a separate piece of paper before going on to the constructions you are given to do in the book.

Many different constructions are explained in this book. After each example, there is an opportunity for you to try one or two similar constructions for yourself. When you try these extra constructions, follow the directions from the example carefully. If you want more practice, you can make up your own angles, segments, shapes to construct and then just follow the instructions in the book to try again.

Many of the constructions that come later in the book involve the same techniques that you will learn in the first few pages. Instead of rewriting all the directions for the basics every time you need the basics in a more advanced construction, you will be referred back to the earlier parts of the book. Make sure that you turn back to those sections of the book if you don't fully remember how to do those basic constructions.

Should you erase all those marks?

A completed construction has dark pencil marks and light compass marks. Sometimes there can be a lot of overlapping arcs and lines. You might have wanted to draw a square and ended up with some of the sides extending past the edges of the square. That's OK. Do not erase the extra marks that are part of doing the construction. If you erase, it looks like a drawing, not a construction. However, it's fine to erase mistakes when you want to try something again.

Helpful hints

1. Once you have measured a distance with a compass, make sure that you hold the compass carefully so that you do not pull the arms farther apart or push them closer together before you do the next step.
2. Do not push too hard on the paper. If you find that your pencil is making a rough dark line, you are pushing the compass down too hard. If you see a big hole in your paper from the compass point, you are pushing too hard on the compassing, or you are pulling on the paper. Be gentle.

3. If your compass seems to fall closed or open easily, tighten the screw where the arms are joined if your compass has one. If it is impossible to tighten the compass arms and your compass will not hold its position when you pick it up, you need a better compass. But before you blame the compass, make sure that you are holding it gently when you use it.

4. If you are having trouble doing part of a construction, try turning the paper so that it faces a different direction. Sometimes just moving the paper can help you to see how that part of the construction looks.

5. Use a separate pencil to draw lines, instead of trying to use the pencil in your compass. You might change the width of your compass when you shouldn't if you try to draw everything with the pencil in the compass.

6. Don't worry if your compass draws over the words in the explanation or if the marks intersect with other parts of your construction. This is not a problem, and it will not affect your construction.

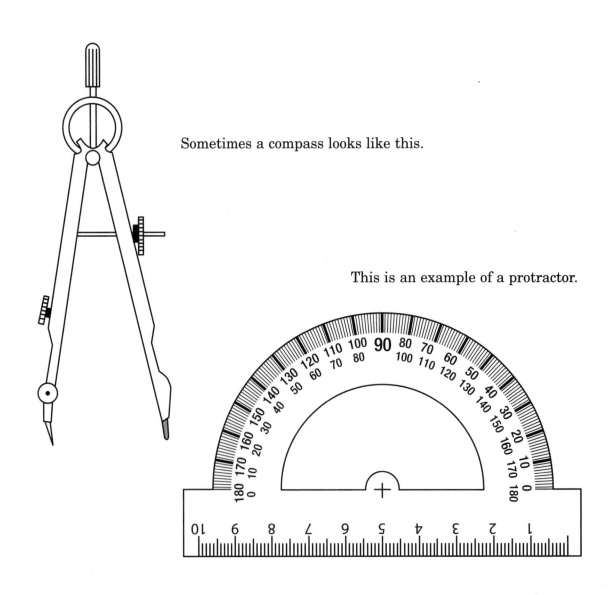

Sometimes a compass looks like this.

This is an example of a protractor.

Chapter 2

Segment Construction

Congruent Segments

Constructing a Congruent Segment When you construct a congruent segment, you are drawing a second segment that has the same length as the given segment.

Here is the given segment. ●────────────●

To construct another segment that is congruent to this one:
1. Draw a line with your straightedge that is longer than the segment you are going to construct. Make one of the endpoints on that new line clear.
2. Measure the length of the original segment by placing the point of segment on one endpoint and stretching the compass so that the pencil reaches the other endpoint.
3. Without changing the compass setting, place the point of the compass on the endpoint of the new line that you drew in Step 1. Make a small arc with the compass which intersects the line you drew.
4. Mark the intersection point. The segment between the two endpoints is congruent to the one you started with.

Try this construction. *Follow the directions above to construct segments congruent to each of these given segments. It does not matter which way the original or new segments face when you do your constructions.*

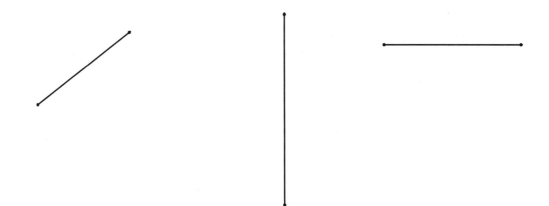

To check your work, try measuring the given segment and your constructed segment with a ruler to see if they have equal lengths.

Segment Additions

In geometry, we are allowed to add the lengths of parts of a segment together to get the total length of the segment. We can demonstrate this idea with constructions.

Example 1: Create a segment which is the length of $\overline{AB} + \overline{CD}$.

1. Draw a line with your straightedge that is longer than the segment you are going to construct. Make one of the endpoints on that new line clear and label it X.
2. Construct a segment congruent to $\overline{AB}$ on the new line. (If you need help doing this, look back at the worksheet on constructing congruent segments.) Label the second endpoint of this segment Y.
3. Construct a segment congruent to $\overline{CD}$ on the new line starting at point Y. Label the new endpoint Z. $\overline{XZ}$ has a length that is equal to the length of $\overline{AB} + \overline{CD}$.

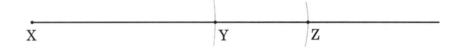

Example 2: Using the given segments above and this additional given segment, construct a segment whose length is equal to the lengths of $\overline{EF} + 2\overline{CD}$ (one of $\overline{EF}$ combined with 2$\overline{CD}$s).

1. First draw a line with your straightedge to work on and label the endpoint L.
2. Construct a segment congruent to $\overline{EF}$ and label the new endpoint M.
3. Construct a segment congruent to $\overline{CD}$ starting at M and label the new endpoint N.
4. Construct a segment congruent to $\overline{CD}$ starting at N and label the new endpoint O. $\overline{LO}$ has a length that is equal to the length of $\overline{EF} + 2\overline{CD}$.

Try this construction. *Follow the directions on the previous page to add segments to create new segments. It does not matter which way the original or new segments face when you do your constructions. All of the given segments are here. You will be asked to combine them to create new segments in each of the exercises.*

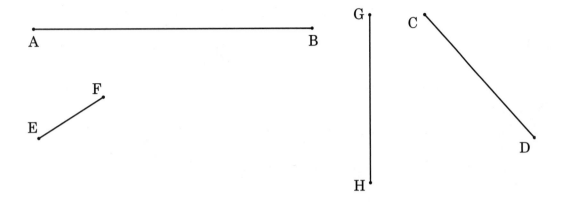

1. Construct a segment whose length is equal to the lengths of $\overline{AB} + \overline{GH}$.

2. Construct a segment whose length is equal to the lengths of $\overline{EF} + \overline{AB}$.

3. Construct a segment whose length is equal to the lengths of $3\overline{EF} + \overline{CD}$.

4. Construct a segment whose length is equal to the lengths of $2\overline{CD} + \overline{AB}$.

5. Construct a segment whose length is equal to the lengths of $\overline{GH} + \overline{EF}$.

Perpendicular Segments

Two segments are perpendicular if they intersect to form right angles. There are two ways to construct a line perpendicular to a given line: 1) through a point on the line; and, 2) through a point not on the line.

Constructing a Perpendicular Segment Through a Point on the Line

1. Pick a point anywhere on the line and label it A. (It helps if the point isn't too close to either end.)

2. Put the point of your compass on point A and make small arcs to the right and left of A which intersect the line. It doesn't matter how wide the compass is, but it needs to stay the same as you make both arcs. Label the two intersection points B and C. (If your arc does not meet the line, either make your compass width smaller and make both arcs again, or choose a different point, closer to the middle of the line, and start over.)

3. Stretch the arms of the compass to make the width of the compass a bit larger. Put the point of your compass on B and make a small arc in the space above point A. Do not change the width of the compass. Put the point of your compass on C and make a small arc in the space above point A that intersects with the arc you just drew from B. (The two arcs should make a little curved X. If they don't meet, make each arc a little longer.)

4. Label the point where the two arcs in step 3 meet D. Connect points A and D with your straightedge. $\overline{AD}$ is perpendicular to the given line.

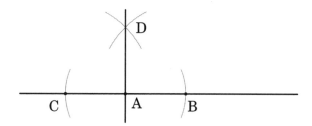

Try this construction. *Follow the directions above to construct two segments that are perpendicular to this one. Make one perpendicular segment through point X and one through point Y. (X and Y are the points in Step 1 of the directions above.)*

To check your work, use a protractor to measure the angles formed by the segment and the given line to see if they are right angles (90°).

Constructing a Perpendicular Segment Through a Point Not on the Line

1. Pick a point anywhere above the line and label it *A*. (It helps if the point isn't too close to either end.)

2. Put the point of your compass on point *A* and stretch the compass straight down until the pencil is on the other side of the line. Then draw an arc like a bowl so that it intersects the line in two places. Label the two intersection points *B* and *C*.

3. Make the compass width slightly smaller. Put the compass point on *B* and make a small arc below the line. Keep the compass width the same. Put the compass point on *C* and make a small arc below the line so that it intersects with the arc you drew from *B*. (The two arcs should make a little curved *X*. If they don't meet, make each arc a little longer. If they still don't meet, increase the compass width and then make <u>both</u> arcs again.)

4. Label the point where the two arcs in Step 3 meet *D*. Connect points *A* and *D*. $\overline{AD}$ is perpendicular to the given line.

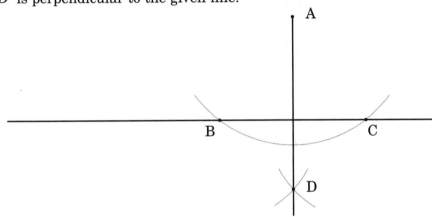

Try this construction. *Follow the directions above to construct two segments that are perpendicular to this one. Make one perpendicular segment through point F and one through point G. (F and G are the points in Step 1 of the directions above.)*

. G

. F

To check your work, use a protractor to measure the angles formed by the segment and the given line to see if they are right angles (90°).

Perpendicular Bisectors

Constructing a Perpendicular Bisector If a line cuts a segment in half and forms a right angle with it, then it is a perpendicular bisector.

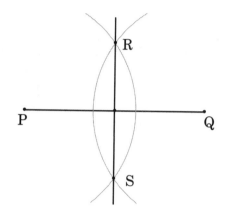

1. Put the point of your compass on point P and stretch the pencil arm until it is between the middle of the segment and point Q. Draw an arc that is a semicircle (half of a circle) extending above and below the segment.

2. Keep the width of the arc exactly the same. Put the point of the compass on Q and make a semicircle like the one you drew in Step 1. This arc needs to intersect on the top and the bottom with the first arc. Label the points of intersection R and S.

3. Draw a segment connecting R and S. $\overline{RS}$ is perpendicular to the given segment and it cuts the segment in half.

Try this construction. *Follow the directions above to construct the perpendicular bisectors of each of these segments.*

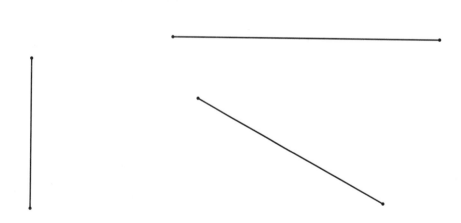

To check your work, use a protractor to measure the angles formed by the segment and the given line to see if they are right angles (90°). Use a ruler to see if the two parts of the given segment are congruent.

Chapter 3

Angle Construction

Congruent Angles

Constructing a Congruent Angle When you construct a congruent angle, you are constructing a second angle that has the same measure as the given angle.

1. Draw a line to be one ray of the new angle. Label one of the endpoints A.

2. Put the point of your compass on point Y of the original angle. Draw an arc that intersects both of the sides of the angle. Label the Y intersection points V and W.

3. Keep the compass width exactly the same. Put the point of the compass on A and draw an arc that looks like the first one you drew. This arc needs to intersect the line you drew in Step 1. Label the intersection point B.

4. Put the point of your compass on V and move the pencil arm until it meets W. When you do this, you are measuring the arc.

5. Keep the compass width the same. Put the point of the compass on B and draw a small arc to intersect the arc you drew in Step 3. Label the point of intersection C.

6. Connect points A and C. This angle is congruent to the one you started with.

Given Angle

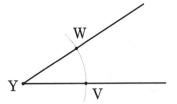

Constructed Angle

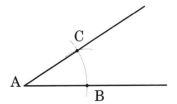

Try this construction. *Follow the directions above to construct angles congruent to these given angles. To check your work, use a protractor to measure both angles.*

Angle Additions

Adding angles is similar to adding segments. You can make bigger angles by combining smaller angles. Use the techniques from constructing congruent angles to build larger angles.

There are two parts to the construction of adding ∠XYZ to ∠ABC.

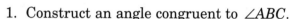

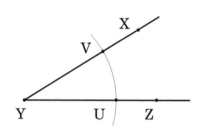

1. Construct an angle congruent to ∠ABC.
   a. Draw a line and label the endpoint Q.
   b. Draw a small arc on ∠ABC, with the point of your compass on B, and label the intersection points D and F.
   c. Draw the same arc on the new line with the point of your compass on Q, and label the intersection point R.
   d. Measure the arc on ∠ABC. Then draw the measurement arc on the arc from R. Label that intersection point W.
   e. Connect Q and W.

2. Use $\overline{QW}$ as the base of the angle congruent to ∠XYZ.
   a. Draw a small arc on ∠XYZ, with the compass point on Y, and label the intersection points U and V.
   b. Draw the same arc on $\overline{QW}$, with the compass point on Q, and label the intersection point T.
   c. Measure the arc on ∠XYZ. Then draw the measurement arc on the arc from T. Label that intersection point S.
   d. Connect S and Q. You now have the sum of the two angles: ∠RQS.

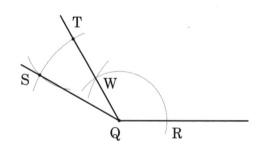

Try this construction. *Here are 3 angles. Do each of the requested angle addition constructions by following the directions on the previous page.*

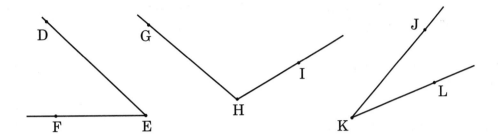

Add ∠DEF to ∠GHI.

Add ∠GHI to ∠JKL.

Angle Bisectors

Constructing an Angle Bisector An angle bisector divides the angle into two congruent angles. The process is the same for acute and obtuse angles, but some of the small hints are different, so there are directions for both an obtuse and an acute angle on this page.

Obtuse Angles

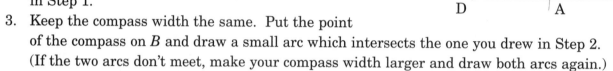

1. Draw an arc which intersects both sides of the angle. Label the intersection points A and B.
2. Make the compass width larger. Put the point of the compass on A and draw a small arc in the interior of the angle outside of the arc you drew in Step 1.
3. Keep the compass width the same. Put the point of the compass on B and draw a small arc which intersects the one you drew in Step 2. (If the two arcs don't meet, make your compass width larger and draw both arcs again.)
4. Label the intersection of the 2 arcs C. Connect points D and C. $\overline{DC}$ is the angle bisector.

Acute Angles

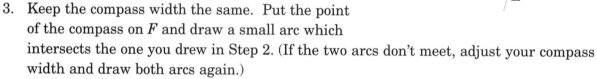

1. Draw an arc which intersects both sides of the angle. Label the intersection points E and F.
2. Make the compass width smaller. Put the point of the compass on E and draw a small arc in the interior of the angle outside of the arc you drew in Step 1.
3. Keep the compass width the same. Put the point of the compass on F and draw a small arc which intersects the one you drew in Step 2. (If the two arcs don't meet, adjust your compass width and draw both arcs again.)
4. Label the intersection of the 2 arcs H. Connect points G and H. $\overline{GH}$ is the angle bisector.

Try this construction. *Construct the bisectors of each of these angles.*

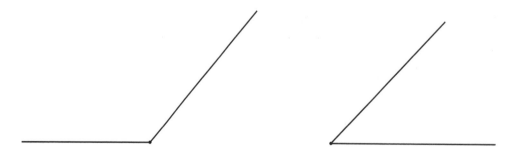

To check your work, use a protractor to measure both small angles you create with the angle bisector.

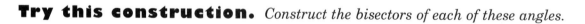

Parallel Lines

Constructing Parallel Lines In order to construct a line parallel to a given line, you need to apply what you learned about constructing congruent angles. The construction relies on the fact that corresponding angles are congruent when there is a pair of parallel lines which are cut by a transversal.

1. Pick a point not on the given line, preferably somewhere below the line, and label it A. (The point can be anywhere, but it will be difficult to do the construction if the point is too close to the line.)

2. Draw a line from the point to the given line. (It doesn't matter how you draw this line as long as it is a straight line which extends beyond both the point and the given line.) Label the intersection with the given line point X.

3. Put the point of your compass on X and draw an arc which intersects both the given line and the line you drew in Step 2. Label the intersection points B and C.

4. Keep the compass width the same. Put the point of the compass on A and draw the same arc facing the same direction. Label the intersection point D.

5. Measure from B to C with your compass. Keep the compass the same width.

6. Put the point of the compass on D and make a small arc intersecting the arc from D. Label the intersection point E.

7. Draw a line that extends through A and E. The line that contains $\overline{AE}$ is parallel to the given line.

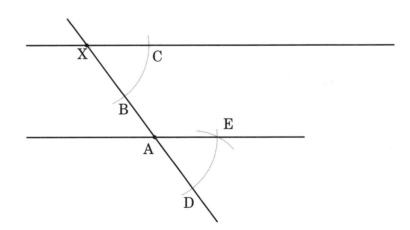

Try this construction. *Construct a line parallel to each of the segments below.*

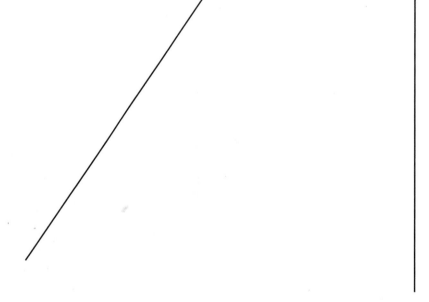

Parallelograms

Constructing a Parallelogram A parallelogram is a 4-sided polygon in which both pairs of opposite sides are parallel. It is also true that both pairs of opposite sides are congruent. These facts, and the following theorem, make this construction possible: If a pair of opposite sides of a 4-sided polygon are congruent and parallel, then the polygon is a parallelogram.

1. Draw a segment to be the top of the parallelogram. Label the endpoints F and G.
2. Pick a point below F that you want to be the lower left-hand corner of the parallelogram. Label it A. Use your straightedge to draw a line that starts at F and continues through A.
3. Construct a line parallel to $\overline{FG}$ through A. (Use the directions from the previous section, *Parallel Lines*, starting with Step 3 to help you.)
4. Construct a segment congruent to segment $\overline{FG}$ starting from A along the parallel line you constructed in Step 3. Label the second endpoint of that segment B.
5. Connect points G and B. You have now constructed parallelogram $FGBA$.

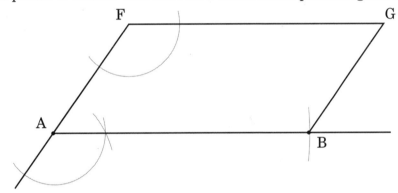

Try this construction. *Construct your own parallelogram here. Use the directions above to help you.*

Rectangles

Constructing a Rectangle A rectangle is a parallelogram with 4 right angles. To do this construction, you will combine what you have learned about constructing a parallelogram with what you previously learned about constructing a perpendicular segment.

1. Draw a segment that is longer than you want the top of the rectangle to be. Label the endpoints E and G. Pick a point between E and G and label it F.

2. Construct a line that is perpendicular to $\overline{EG}$ through point F. (Look back at how to construct a perpendicular segment through a point on the line if you need help.)

3. Pick a point A on the perpendicular segment to be the lower left-hand corner of your rectangle. Construct a line perpendicular to $\overline{FA}$ through point A. (This second perpendicular segment is parallel to the segment you drew in Step 1.)

4. Construct a segment congruent to $\overline{FG}$ starting from A along the perpendicular line you constructed in Step 3. Label the second endpoint of that segment B.

5. Connect points G and B. You have now constructed rectangle $FGBA$.

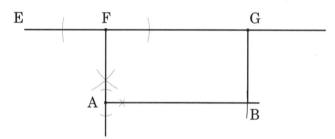

Try this construction. *Construct your own rectangle below. Follow the directions above.*

Squares

Constructing a Square A square is a rectangle with 4 congruent sides. The process of constructing a square is nearly the same as constructing a rectangle.

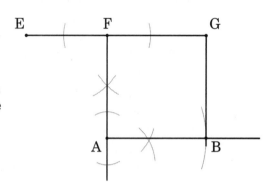

1. Draw a segment that is longer than you want the top of the square to be. Label the endpoints E and G. Pick a point between E and G and label it F.

2. Construct a line that is perpendicular to $\overline{EG}$ through point F. (Look back at how to construct a perpendicular segment through a point on the line if you need help.)

3. Measure the length of $\overline{FG}$ with your compass. Construct a segment congruent to $\overline{FG}$ on the perpendicular segment you constructed in Step 2. Label the new endpoint A.

4. Construct a line perpendicular to $\overline{FA}$ through point A. (This second perpendicular segment is parallel to the segment you drew in Step 1.)

5. Construct a segment congruent to $\overline{FG}$ starting from A along the perpendicular line you constructed in Step 4. Label the second endpoint of that segment B.

6. Connect points G and B. You have now constructed square $FGBA$.

Try this construction. *Construct a square by following the directions above.*

4 5 ° A n g l e s

Constructing a 45° angle You have already learned how to construct a perpendicular segment and to use that construction to make a variety of rectangles. Now you will combine the perpendicular line construction with the angle bisector construction. Since perpendicular lines form right angles (90°), bisecting that right angle will result in two 45° angles.

1. The simplest way to create a right angle is to construct a perpendicular bisector, so construct a perpendicular bisector of this segment. (Look back to those instructions on page 12 if you need help.) $\overline{AB}$ is the perpendicular bisector constructed here.

2. Construct an angle bisector of one of the right angles formed by the perpendicular segment. (Look back to those instructions if you need help.) Both $\angle ACD$ and $\angle DCE$ are 45° angles.

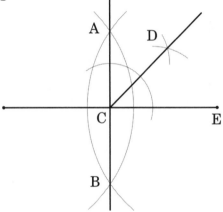

To check your work, use a protractor to measure the small angle you created. It should be 45°.

Try this construction. *Follow the directions above to create another 45° angle.*

Right Triangles

Constructing a Right Triangle A right triangle has two legs which intersect to form a right angle. The third side is called the hypotenuse, so you can apply the perpendicular segment constructions to construct a right triangle.

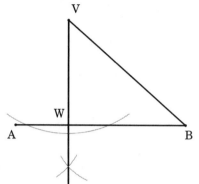

1. Draw a segment and label the endpoints *A* and *B*. Choose a point above that segment to be a vertex of your right triangle and label that point *V*.

2. Construct a segment perpendicular to the first segment through *V*. (Look back at the instructions on page 10 for how to construct a perpendicular segment through a point not on the line for help.) Label the intersection point *W*.

3. Use your straightedge to connect *V* and *B*. You have constructed right triangle △*VWB*.

Give this construction a try. *Use the instructions above to construct another right triangle.*

E q u i l a t e r a l   T r i a n g l e s

Constructing an Equilateral Triangle An equilateral triangle has 3 congruent sides and three congruent angles which each measure 60°.

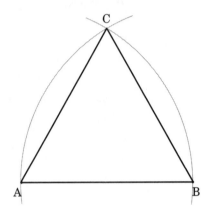

1. Draw a segment to be the bottom of your equilateral triangle. Label the endpoints *A* and *B*.
2. Put the point of the compass on *A* and stretch the compass until the pencil is on *B*. Draw an arc from *B* until the arc ends above *A*.
3. Keep the compass width the same. Put the point of your compass on *B* and draw an arc from *A* up to intersect with the first arc you drew. Label the intersection point *C*.
4. Connect *C* to *A* and *B*. You have constructed an equilateral triangle.

To check your work, use a ruler to measure all three sides. They should all be the same length.

Try this construction again. *Start with a different length segment and follow the instructions above to create another equilateral triangle.*

30° Angles

Constructing a 30° Angle Since an equilateral triangle has 3 angles, each having a measure of 60°, the construction of an equilateral triangle is the first step in constructing a 30° angle. This is combined with the construction of a congruent angle and the construction of an angle bisector.

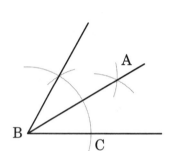

1. Construct an equilateral triangle. (Look back at the instructions.)
2. Construct an angle congruent to one of the 3 angles.
3. Construct the angle bisector of the angle you created in Step 2. (For Step 1 of constructing an angle bisector, you can use the arc you drew to create the congruent angle.) You now have a pair of 30° angles, one of which is ∠ABC.

To check your work, use a protractor to measure the small angles you created to see if they each measure 30°.

Try this construction again. *Construct a 30° angle by following the directions above. For a shortcut, use an equilateral triangle you constructed previously.*

I s o s c e l e s   T r i a n g l e s

Constructing an Isosceles Triangle An isosceles triangle has 2 congruent sides. The two congruent sides are called the legs. The side that is not congruent is called the base. The construction of an isosceles triangle is similar to the construction for an equilateral triangle.

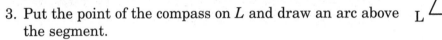
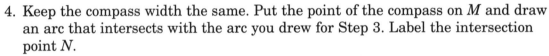

1. Draw a segment to be the base of your isosceles triangle. Label the endpoints L and M.
2. Measure the base with your compass. Then stretch the compass so that the width of the compass is longer than the base.
3. Put the point of the compass on L and draw an arc above the segment.
4. Keep the compass width the same. Put the point of the compass on M and draw an arc that intersects with the arc you drew for Step 3. Label the intersection point N.
5. Connect N to L and M. You have constructed an isosceles triangle.

To check your work, use a ruler to measure the legs to see if they are congruent.

Try this construction again. *Follow the directions above to construct another isosceles triangle. Try to make this triangle different from the first one you made.*

Chapter 4

Constructions Based on Congruent Triangle Theorems

The next 3 constructions are based on theorems that are learned in geometry class to prove that two triangles are congruent. The theorems used in these constructions are the ASA (Angle Side Angle) and SAS (Side Angle Side) theorems. The constructions of the equilateral and isosceles triangles (refer to Chapter 3) are based on the SSS (Side Side Side) theorem.

A Triangle from 2 Given Angles and 1 Given Segment

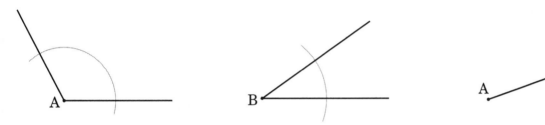

1. Construct a segment congruent to the given segment, $\overline{AB}$.
2. Construct an angle congruent to the first given angle with its vertex, A, on the left endpoint of the segment, also A.
3. Construct an angle congruent to the second given angle with its vertex, B, on the right endpoint of the segment, also B. You will have to reverse the direction that the angle opens.
4. If the sides from the angles don't meet, extend them until they do. Label the intersection point C. You have constructed a triangle from these parts.

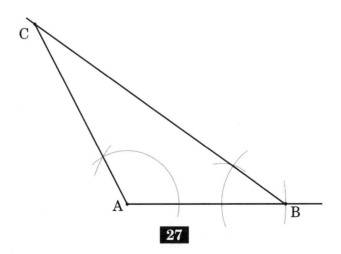

Try this construction. *Construct a triangle from these parts.*

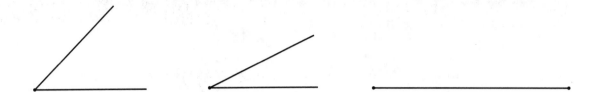

A Triangle from 2 Given Segments and 1 Given Angle

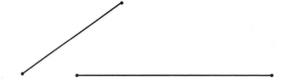

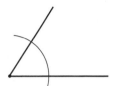

1. Construct a segment congruent to the first given segment. Label the endpoints *A* and *B*.

2. Construct an angle congruent to the given angle with its vertex at *A*.

3. Construct a segment congruent to the second given segment starting at *A* along the second side of the angle you constructed. Label the second endpoint of that segment *C*.

4. Connect *C* and *B*. You have constructed a triangle from these parts.

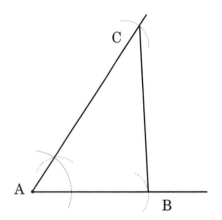

Try this construction. *Construct a triangle from these parts.*

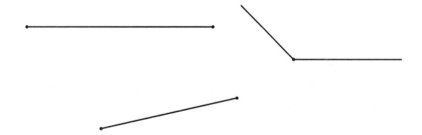

A Triangle Congruent to a Given Triangle

This construction combines constructing congruent segments, constructing congruent angles, and using the same basic approach as the last construction you learned (constructing a triangle from 2 given segments and a given angle).

1. Choose a side of the triangle and construct a segment congruent to that segment. (In the example, a segment was constructed which is congruent to $\overline{AB}$.)

2. Construct an angle congruent to one of the angles which has $\overline{AB}$ as a side of the angle. (In the example, the angle that was constructed is congruent to $\angle ABC$.)

3. From the vertex of the angle you just constructed, construct a segment congruent to $\overline{BC}$ along the second side of the angle.

4. Connect the endpoint you just constructed to the endpoint of the other segment of your triangle. You have now constructed a triangle congruent to the given triangle.

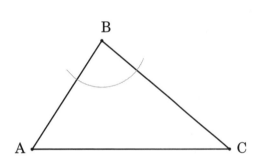

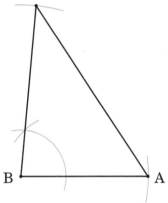

Construct a triangle congruent to this one. Follow the directions above.

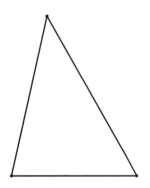

Chapter 5

Special Segments in Triangles

Each of the constructions in this section involves the basic constructions you have been using throughout this book. The median, angle bisectors, perpendicular bisectors and altitudes are unique because they are important elements of triangles in the study of geometry.

Angle Bisectors

There are 3 angles in a triangle, one at each vertex. Therefore, there are three angle bisectors in every triangle.

To construct the angle bisector for the angle whose vertex is at A:
1. Ignore side $\overline{BC}$. Construct the angle bisector for $\angle BAC$ as you learned previously.

Try this construction. *Construct the angle bisectors of the other 2 angles of this triangle.*

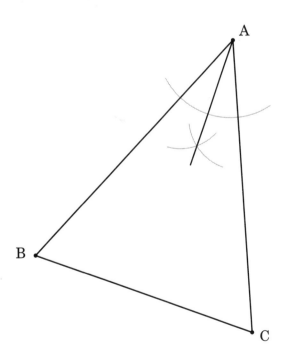

Perpendicular Bisectors

There are 3 perpendicular bisectors in a triangle, one on each side of the triangle.

To construct the perpendicular bisector for side $\overline{DE}$.:
1. Ignore the other two sides of the triangle and treat $\overline{DE}$ as if it were a segment alone. Construct the perpendicular bisector of the segment as you learned previously.

Try this construction. *Construct the perpendicular bisectors of the other 2 sides of this triangle.*

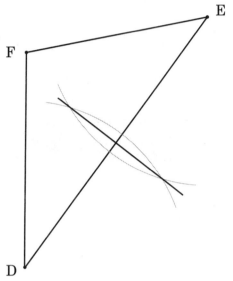

Medians

The median of a triangle connects a vertex to the midpoint of the opposite side. Therefore, you start by finding the midpoint of the side, and then you use your straightedge to connect the midpoint to the vertex.

To construct the median from *X:*
1. Find the midpoint of $\overline{YZ}$ by constructing the perpendicular bisector of $\overline{YZ}$. (The midpoint of $\overline{YZ}$ is the point where the bisector and the segment intersect. Label the midpoint *W.*)
2. Use the straightedge to connect *X* and *W.* $\overline{XW}$ is the median from *X.*

Try this construction. *Construct the other 2 medians in this triangle.*

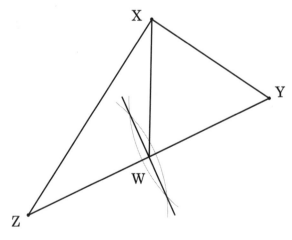

32

Altitudes

An altitude goes from a vertex to the opposite side, and it is perpendicular to the opposite side.

Altitudes in Acute Triangles

To construct the altitude from G to $\overline{HI}$:

1. Use the construction for constructing a perpendicular segment through a point not on the line. G is the point not on the line and $\overline{HI}$ is the line.
2. Label the intersection point J. $\overline{GJ}$ is the altitude from G.

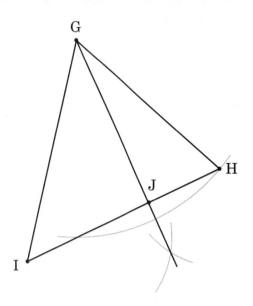

Construct the 2 other altitudes in this triangle.

Altitudes in Obtuse Triangles

To construct the altitude from K to $\overline{JL}$:

1. Use the construction for constructing a perpendicular segment through a point not on the line. However, for two of the altitudes, the perpendicular segment will not reach the side of the triangle. So you need to begin this construction by extending the sides of the triangle beyond the obtuse angle. In this example, K is the point not on the line and the extended $\overline{JL}$ is the line.
2. Label the intersection point M. $\overline{KM}$ is the altitude from K.

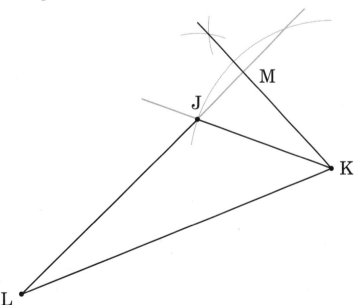

Construct the 2 other altitudes in this triangle. *Hint: One of the altitudes will be like the example above. The altitude from the obtuse angle will be like the example for the altitude in an acute triangle.*

Chapter 6

Circle Constructions

Tangents to a Circle

Constructing a Tangent to a Circle A tangent to a circle is a line that intersects the circle at only one point. A tangent is perpendicular to the radius that goes from the center of the circle to the point of tangency. Since the radius and tangent are perpendicular, you can use the construction for a perpendicular segment through a point on the line.

1. Pick a point on the circle and label it B.
2. Use your straightedge to draw a line which starts at Point A, the center of this circle, and continues through B.
3. Do the construction of a perpendicular segment through a point on the line. B is the point and $\overline{AB}$ is the line. The perpendicular line you construct is the tangent to this circle.

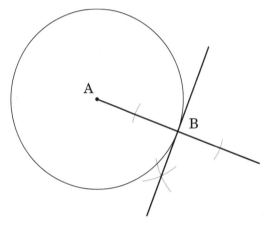

Try this construction. *Construct tangents to these circles.*

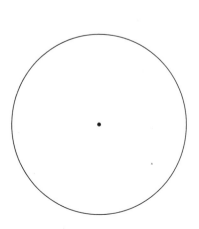

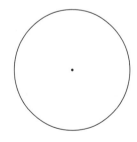

A Circle Circumscribing a Triangle

Constructing a Circle Which Circumscribes a Triangle When a circle is circumscribed about a triangle, each of the vertices of the triangle meets the circle in a point of tangency. The simplest method of constructing this type of figure is to use the construction for the perpendicular bisectors of the sides of the triangle. It is extremely important to hold the compass delicately and to be precise when doing circle constructions.

1. Construct the perpendicular bisector of one of the sides of the triangle.
2. Construct the perpendicular bisector of another side of the triangle. (It does not matter which two sides you pick for Steps 1 and 2.)
3. Label the point of intersection of these two perpendicular bisectors X.
4. Put the point of your compass on X. Stretch the arms of your compass apart until the pencil point is resting on one of the vertices of the triangle. (It does not matter which vertex you choose.)
5. Keep this compass width steady and turn the compass to construct a circle whose center is the point X. You have circumscribed a circle about this triangle.

Try this construction. *Circumscribe circles about each of these triangles.*

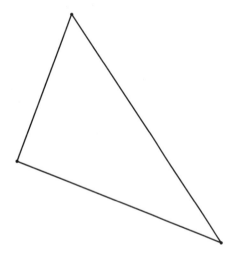

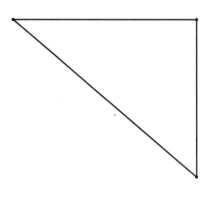

A Circle Inscribed in a Triangle

Constructing a Circle Inscribed in a Triangle When a circle is inscribed in a triangle, each of the sides of the triangle is a tangent to the circle; the circle is completely inside the triangle and touches the triangle at only 3 points. It is extremely important to hold the compass delicately and to be precise when doing circle constructions.

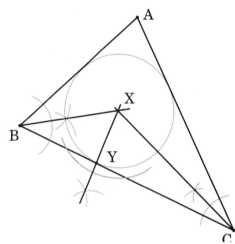

1. Construct an angle bisector on one of the angles of the triangle.
2. Construct an angle bisector on another of the angles of the triangle. (It does not matter which two angles you choose for Steps 1 and 2.)
3. Label the point of intersection of these two angle bisectors X.
4. Construct a perpendicular segment from X to the segment which joins the two angles you bisected. (In the example, $\angle ABC$ and $\angle BCA$ were bisected, so the perpendicular from X to $\overline{BC}$ was constructed.) Use the construction for a perpendicular segment to a line from a point not on the line.
5. Label the point of intersection of the perpendicular segment and $\overline{BC}$ point Y.
6. Put the point of your compass on X and stretch the arms apart until the pencil point is resting on Y. Keep this compass width steady and turn the compass to construct a circle whose center is the point X. You have inscribed a circle in this triangle.

Try this construction. *Inscribe circles in each of these triangles.*

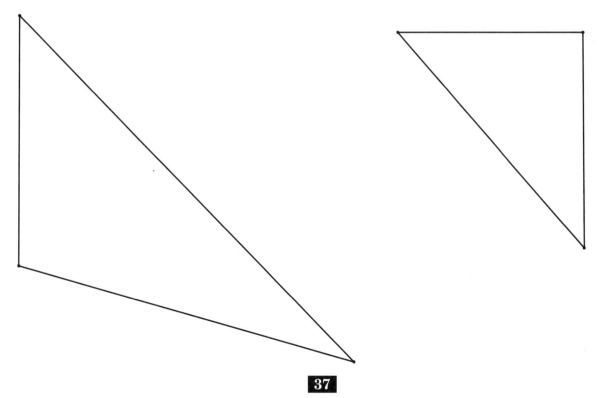

Constructions Exam

Directions: *This exam requires you to put together everything you have learned in this book. Look at each question and think about how you can break it down into constructions you have learned how to do before you start to answer.*

1. Construct an angle that has a measure of 75°.

2. Circumscribe a circle about this triangle.

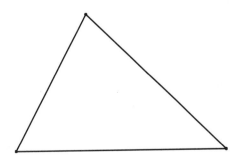

3. Construct an isosceles triangle and construct an altitude.

4. Construct a parallelogram.

5. Inscribe a circle in this triangle.

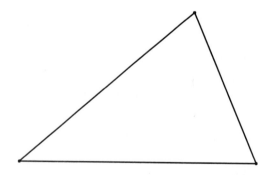

Glossary

Angle: Two rays that start from the same endpoint.

Arc: A part of a circle.

Auxiliary Line: A line that is added to a picture. It is sometimes helpful to extend a line or add the ray of an angle to solve a geometric problem. Auxiliary lines are usually dashed so that it is clear that they aren't part of the original problem.

Complementary Angles: Two angles whose sum is 90°.

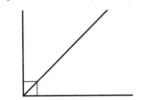
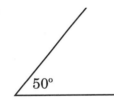
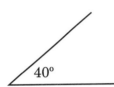

Parallel Lines: Two lines which do not intersect.
- $\overline{AC} \| \overline{DF}$ is read as "line AC is parallel to line DF."
- $\overline{GH}$ is a transversal because it intersects 2 lines.
- The arrows on the lines show that they are parallel.
- When 2 parallel lines are cut by a transversal
  a. Alternate interior angles are congruent,
     $\angle CBE \cong \angle BED$ and $\angle ABE \cong \angle BEF$.
  b. Corresponding angles are congruent $\angle CBA \cong \angle BED$
     and $\angle GBC \cong \angle BEF$ and $\angle ABE \cong \angle DEH$
     and $\angle CBE \cong \angle FEH$.
  c. Same-side interior angles are supplementary,
     $m\angle ABE + m\angle DEB = 180°$ and $m\angle CBE + m\angle FEB = 180°$.

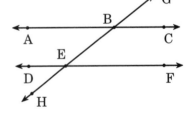

Parallelogram: A quadrilateral in which both pairs of opposite sides are parallel.
- Pairs of opposite sides are congruent.
- Pairs of opposite angles are congruent.
- Diagonals bisect each other.

Perpendicular Lines: Two lines which intersect to form right angles.

Radius: A straight line extending from the center of a circle.

Rectangle: A parallelogram with 4 right angles.
- All of the properties of parallelograms also apply to rectangles.
- Diagonals are congruent to each other.

Square: A rectangle with 4 congruent sides.
- All properties of parallelograms and rectangles apply to squares.
- Diagonals intersect to form right angles.
- Diagonals bisect the angles of the parallelogram.

Supplementary Angles:
- Two angles whose sum is 180°.

- If two adjacent angles form a straight line, they are supplementary.

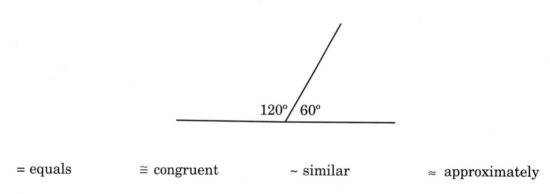

| = equals | ≅ congruent | ~ similar | ≈ approximately |

Tangent: A straight line perpendicular to the radius of a circle.

Vertex: The point farthest from the base of a triangle.

Answer Key

This key is a guide to give you a sense of whether or not you understand how to do these constructions. Since there are many individual decisions to be made when doing constructions, your work can be correct without looking exactly like the constructions done here. This is a limited answer key—not all of the exercises have been completed here.

Page 7: Congruent Segments

To check your work, measure each of the segments you constructed and the segments that were given. If they are the same, you did the construction correctly. Your constructed segment does not have to face the same direction as the given segment; it just has to be the same length.

Page 9: Segment Additions

Your completed segments should have the lengths shown here. The work has not been shown so that you can focus on the final length for each exercise.

1. ●————————————————————————————●

2. ●————————————————————————●

3. ●—————————————————————————●

4. ●——●

5. ●————————————————●

Page 10: Perpendicular Segments

The best way to check whether you have done this construction correctly is to measure it with a protractor. It should be a 90° angle. The construction marks from the example should be present in your construction.

Page 11: Perpendicular Segments

The best way to check whether you have done this construction correctly is to measure it with a protractor. It should be a 90° angle. The construction marks from the example should be present in your construction.

Page 12: Perpendicular Bisectors

The best way to check whether you have done this construction correctly is to measure it with a protractor. It should be a 90° angle. Use a ruler to check whether the two small segments you have created on the given segment are of equal length. The construction marks from the example should be present in your construction.

Page 13: Congruent Angles

To check your work, use a protractor to measure each of the angles you constructed and the angles that were given. If they are the same, you did the construction correctly. Your constructed angle does not have to face the same direction as the given angle; it just has to have the same number of degrees.

Page 15: Angle Additions

For each of these angle addition constructions, your picture doesn't have to face the same way as this example, and your construction marks might be in a different place. But if you measure **the** final angle and **your** final angle, they must have the same measure.

Add ∠DEF to ∠GHI.

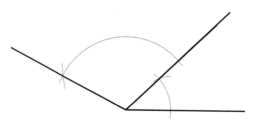

Add ∠GHI to ∠JKL.

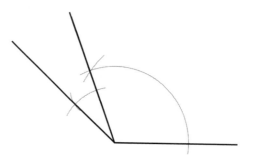

Page 16: Angle Bisectors

To check your angle bisectors, measure each of the smaller angles as has been done here. If the two smaller angles have the same measure, then the construction is correct.

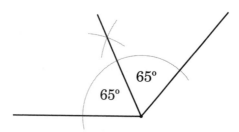

65° 65°

Page 18: Parallel Lines

Your construction should be nearly identical to the example. You can measure the distance between the given line and your constructed line in two places (make sure to hold the ruler perpendicular to the two lines both times). If the two measurements are the same, then the lines are parallel.

Page 19: Parallelograms

Every parallelogram will be different. But you can check to see that the opposite sides are of equal length. As long as the top and bottom are congruent, and the sides are congruent, then you know that you have a parallelogram.

Page 20: Rectangles

You can check your rectangle by first making sure that opposite sides are congruent. Then measure the angles in the corners with a protractor. They should be 90° angles.

Page 21: Squares

You can check your square by first making sure that all sides are congruent. Then, measure the angle in the corners with a protractor. They should be 90° angles.

Page 22: 45° Angles

This is a complicated construction. Here it is broken into steps.

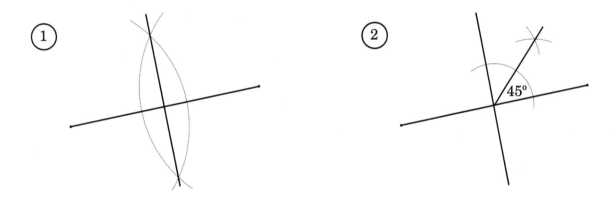

Page 23: Right Triangles

Your triangle should have the same markings as the example, though it can be facing in any direction. You can check your work by measuring $\angle VWB$ to make sure that it measures 90°.

Page 24: Equilateral Triangles

Equilateral Triangle: Measure all three sides of your triangle. If they are all the same, then you have done the construction correctly. When you measure each of the small angles in the bisected angle with a protractor, they should both measure 30°.

Page 25: 30° Angles

If the two legs of your triangle have the same measure, then you have done the construction correctly.

Page 26: Isosceles Triangles

You can check your isosceles traingle by first making sure that two of the sides are congruent. Them meausre the angles with a protractor. Two angles should have the same measure.

Page 28: Congruent Triangles: 2 Angles/1 Segment

Your triangle might face in a different direction, but it should be the same size and shape as this one.

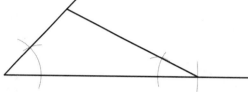

Page 29: Congruent Triangles: 2 Segments/1 Angle

Your triangle might face in a different direction, but it should be the same size and shape as this one.

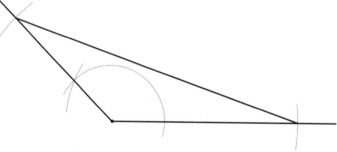

Page 30: Congruent Triangles: Triangle to Trinangle

If you measure all three sides of the triangle you constructed, they should have the same lengths as the sides of the given triangle.

Pages 31–33: Special segments in Triangles

The triangles should look like these with the other angle bisectors and perpendicular bisectors.

Page 31: Angle Bisectors

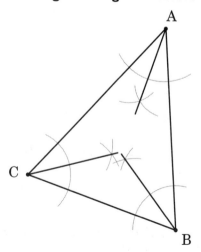

Page 32: Perpendicular Bisectors

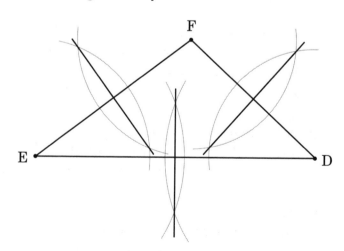

Page 32: Medians

Page 33: Altitute: Acute Triangles

The triangles should look like this with the other medians and altitudes.

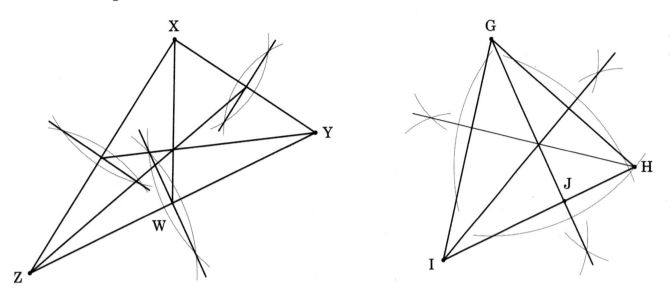

Page 34: Altitude: Obtuse Trinangles

The triangle should look like this with the other altitudes.

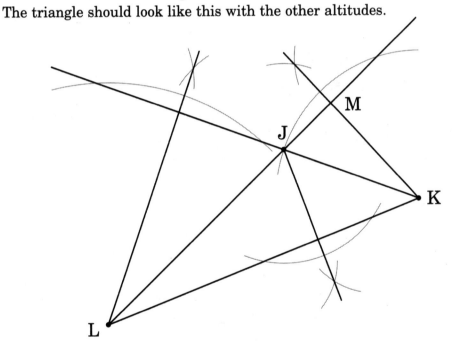

Page 35: Tangents to a Circle

To check that you have constructed the tangent correctly, measure the angle between the radius and the tangent with a protractor to see that it is a 90° angle.

Pages 36–37

You can have problems with these constructions. It is very important that you use a sharp pencil in your compass and that you do the construction carefully. The circle should touch the triangle at the vertices.

Page 39–40 Constructions Exam

I. Construct an angle that has a measure of 75°.

   a. Construct an equilateral triangle.

   b. Construct a 30° angle.

   c. Construct a perpendicular segment to a line.

   d. Bisect the right angle to construct a 45° angle.

   e. Add the 45° angle and the 30° angle to construct the 75° angle.

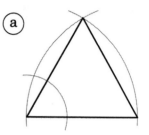

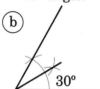

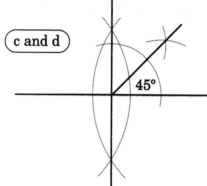

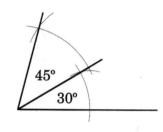

2. Circumscribe a circle about this triangle.

   a. Construct the perpendicular bisectors of 2 sides.

   b. Use the point of intersection as the center for your circle and the distance to one of the vertices as the radius.

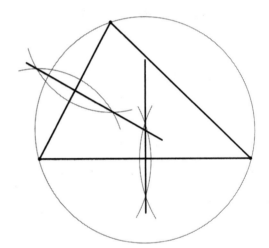

3. Construct an isosceles triangle and construct an altitude.

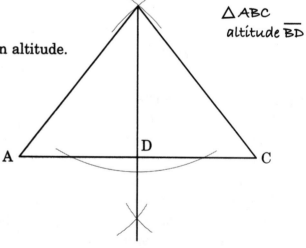

△ ABC
altitude $\overline{BD}$

4. Construct a parallelogram.

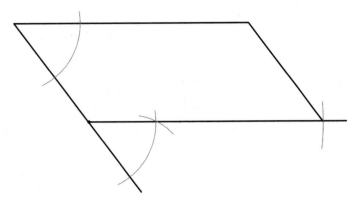

5. Inscribe a circle in this triangle.

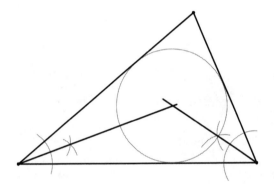

GarlicPress
Tools for Learning and Growing

Math Series

The Straight Forward Math Series

is systematic, first diagnosing skill levels, then *practice*, periodic *review*, and *testing*.

Blackline

GP-006 Addition
GP-012 Subtraction
GP-007 Multiplication
GP-013 Division
GP-039 Fractions
GP-083 Word Problems, Book 1
GP-042 Word Problems, Book 2

The Advanced Straight Forward Math Series

is a **higher level** system to **diagnose**, practice, review, and test skills.

Blackline

GP-015 Advanced Addition
GP-016 Advanced Subtraction
GP-017 Advanced Multiplication
GP-018 Advanced Division
GP-020 Advanced Decimals
GP-021 Advanced Fractions
GP-044 Mastery Tests
GP-025 Percent
GP-028 Pre-Algebra, Book 1
GP-029 Pre-Algebra, Book 2
GP-030 Pre-Geometry, Book 1
GP-031 Pre-Geometry, Book 2

Upper Level Math Series

GP-104 Algebra, Book 1
GP-105 Algebra, Book 2
GP-045 Trigonometry
GP-054 Geometry
GP-053 Pre-Calculus
GP-064 Calculus AB, Vol. 1
GP-067 Calculus AB, Vol. 2

2 SIDED
Self-Checking Math Puzzles

Each puzzle set contains 10 individual puzzles. Each six-inch puzzle is two-sided. One side contains basic math facts, the other side has a photograph. Each puzzle has its own clear plastic tray and lid.

Math problems are solved in the bottom tray (answer pieces are all the same shape). The lid is closed and the puzzle is turned over. If the photo is jumbled, the math facts have not been completed correctly.

GP-113 Addition Puzzles
GP-114 Subtraction Puzzles
GP-115 Multiplication Puzzles
GP-116 Division Puzzles
GP-122 Multiplication & Division Puzzles
GP-123 Money Puzzles

front puzzle   back photo